Anonymous

Health Resorts of the Salt River Valley in Arizona

including Prescott, Jerome and Castle Creek Hot Springs

Anonymous

Health Resorts of the Salt River Valley in Arizona
including Prescott, Jerome and Castle Creek Hot Springs

ISBN/EAN: 9783744772969

Printed in Europe, USA, Canada, Australia, Japan

Cover: Foto ©berggeist007 / pixelio.de

More available books at **www.hansebooks.com**

THE HOTEL ADAMS, PHŒNIX.

Health Resorts

of the

Salt River Valley

in Arizona,

Including Prescott, Jerome and Castle Creek Hot Springs.

Issued by the Passenger Department,
Santa Fe Route,
1898.

The Benefits of Climate Cure.

ARE your lungs weak? Have you a troublesome cough? Is your throat disagreeably affected by each change of weather? Are your nerves unstrung?

If thus afflicted, try the "climate cure," which consists of taking up one's residence in a place where the air is dry, pure, warm and equable, and where proper medical attendance and nursing may be had if needed. Nature is the best of physicians when conditions are friendly.

Where shall one go for friendly conditions? The best place in this country, and perhaps the best in the world, is the Rocky Mountain region of Colorado, New Mexico, Arizona and California. Some localities in that immense uplift of the earth's crust are better than others. The Salt River Valley, in south-central Arizona, takes high rank among the districts where lung and throat troubles can be relieved.

If you doubt the value of "climate cure," or the climatic advantages of the southwestern part of the United States, please glance over the following extracts from a recent article in the Kansas City *Medical Record*, written by Dr. L. W. Luscher:

"In proposing a change to a special climate, I do not propose to exclude the use of any of the so-called specific remedies that have proven beneficial in the hands of physicians, nor any general remedies that could in any way help to support and sustain the patient; in fact, I would advise active treatment at the hands of experienced physicians.

"In selecting a climate for such a patient there are several things to be considered. To begin with, it should be so isolated by altitude or protected by mountains as to remove it from the influence of quick changes in temperature and of the barometer. The air and surroundings generally should be as free as possible from pyogenic and putrefactive bacteria, and as undiluted sunlight is one of the greatest enemies of bacteria, there should be as many days of sunshine in the year as can be obtained.

"Your location should be protected from sudden changes either by altitude or, what is better, mountain ranges. The air should be dry, with little or no rainfall, having as many clear days as possible. The temperature should be such that the patient could spend the entire day in the open air and sunlight. The mountains have advantages, also great disadvantages. The air is purer as you ascend, owing to the decrease of moisture and the consequent relative absence of pyogenic and putrefactive bacteria;

hence the patient improves more rapidly as he ascends into purer air.

"Cold of the higher altitudes prevents out-of-door life, while others can never reach the high altitudes, owing to the rarity of the air. A lower altitude with the same pure air would be better for all; besides, the mountain is lacking in sunshine.

"The English go to Egypt, and the French to Algiers. There they have a warmer climate and more sunshine, but the atmosphere is humid, and bacteria thrive. It is only in our own North America that almost ideal conditions are obtained: We have in the West, between the two great ranges of the Rocky Mountains, a great plateau extending from the 104th to the 112th meridian west, and from the 32d to the 42d parallel north to the southern boundary of the United States, and along the same plateau into Mexico for some distance. In the United States it includes a large part of Colorado and Utah, with New Mexico, Arizona and a part of western Texas.

"In any part of this vast territory the consumptive can find almost all the advantages required: altitude ranging from a few hundred feet up to ten thousand, a freedom from quick changes of temperature and barometric pressure — especially is this the case in the southern part; a very light rainfall, a very low average atmospheric humidity, ranging from year to year 30 to 40 per cent on a scale of 100; as to sunshine, they have an average of 235 perfectly clear days, with 99 days partly cloudy, and only 31 days when the sky is overcast the entire day.

"At almost any part of this vast territory even a sick man could spend 325 days of the year in the open air. Lastly, and what is perhaps of the most importance, there are no putrefactive and almost no pyogenic bacteria found. Wounds heal in the open air without pus, while fresh meat cures without salt when hanging in the open air with hot sun rays pouring upon it. Dead animals on the plains do not putrefy, but dry up. There is no odor produced by them.

"Here, it appears to me, we have the nearest possible approach to an ideal climate for consumptives: altitudes to suit any kind of heart; temperature of low mean variation, and tempered to suit the most delicate organization; a maximum of sunshiny days; a minimum of moisture, with almost entire absence of disease-producing bacteria; air that is pure without being too much rarified.

"It is, of course, understood that climate is not a specific for pulmonary tuberculosis, but that with such favorable surroundings, assisted by judicious treatment, Nature is able to bring about a cure when under less favorable surroundings a cure would be improbable.

"Theoretically, residence on the great American plateau promises the best for persons suffering from pulmonary tuberculosis. Practically, many thousands of residents in robust health, whose lives had been despaired of before going there, give proof of the validity of the promise."

The Persia of America.

"Charles Dudley Warner gave a new distinction to Southern California when he called it 'Our Italy.' This phrase suggests another which may be aptly applied to Arizona. If Southern California is our Italy, Arizona is our Persia, in soil, in climate, in productions and in the character of its landscape. It is much more like Persia than it is like any other locality in the United States, and in the next ten years it may well show the world what Persia might have been

AN IRRIGATING CANAL.

about the dawn of the twentieth century if it had fallen into the hands of the Anglo-Saxon. A good way to describe Arizona briefly is to put before the reader the following sketch of Persia from a popular cyclopedia:

"'The climate is very dry everywhere in the country, except in the Caspian coastlands. In the valleys it is hot, with mild winters. On account of the dryness, both of the climate and the soil, the country bears in many places a naked and barren aspect, but wherever sufficient water can be procured, and irrigation is carried on, the life of Nature develops immediately into a fairy tale.

"'Persia is the home of the rose and the nightingale. In the valleys, the cypress and myrtle abound, the fig grows wild, the mulberry and olive are cultivated in large plantations, the vineyards yield strong and highly flavored wines; apples, pears, apricots, peaches, cherries, oranges and pomegranates of unsurpassed quality are raised in the orchards, and the gardens teem with roses and geraniums. The date-palm grows in the oases of the desert, and dates are a common article of food. The cereals are wheat of excellent quality, rice, maize and barley. It is characteristic of Persia, for its climate and soil, not for its method or energy of cultivation, that many of the fruits which it produces are unequaled in nourishing power, in savoriness, in richness of flavor, and in beauty of appearance, by any of the same kind produced elsewhere on the earth.'

"This is Persia. This is also Arizona. In soil and climate, in the range of production and in physical aspect the one is the counterpart of the other. But Persia is sleeping peacefully, in a neglected corner of Asia Minor, while Arizona is on the broad highway of American civilization and wide-awake to her opportunities. There is but one reason why it is worth while to call

Southern California 'Our Italy,' or Arizona 'Our Persia.' This is because the average American citizen knows much more about foreign lands than about the new empires that are being developed in his own country, so that the shortest route to his understanding is to tell him that California has the soft climate and semi-tropical luxury of Italy, and that Arizona has the dry air, even temperature and marvelous productiveness of Persia when her deserts are overcome by irrigation."—*Irrigation Age.*

Arizona: A Winter Resort.

Hon. Whitelaw Reid, editor of the New York *Tribune*, and formerly United States Minister to France, spends nearly every winter at Phœnix, Arizona, where, for the sake of greater comfort, he rents a house and lives in a home-like fashion.

Mr. Reid's article on this country, reprinted below, which originally appeared in the editorial columns of the *Tribune*, under date of November 22, 1896, will be read with interest:

"So many questions are asked about Arizona as a place for winter residence, and there appears to be such a dearth of precise information among many who are vitally interested, that it seems almost a public duty to set down, in the simplest form, a few facts of personal observation.

"WEATHER.— During a five months' residence in Southern Arizona, in winter, there was but one day when the weather made it actually unpleasant for me to take exercise in the open air at some time or other during the day. Of course there were a good many days which a weather observer would describe as 'cloudy,' and some that were 'showery'; but during these five months (from November, 1895, to May, 1896,) there were only four days when we did not have brilliant sunshine at some time during the day. Even more than Egypt, anywhere north of Luxor, Arizona is the land of sunshine. As to details :

"TEMPERATURE. — I have seen the thermometer mark 92° in the shade on my north piazza in March. On the other hand, we had frosts which killed young orange trees, and there were several nights when thin ice formed. The Government reports show a mean temperature for fourteen years at the present Territorial capital of 57½° in November, 53° in December, 49° in January, 54° in February, 61° in March and 66° in April. The same reports show the highest and lowest temperatures, averaged for eight years, at the same place, as follows: For November 78½° and 42°, December 73½° and 36½°, January 65½° and 32°, February 71½° and 35½°, March 81½° and 41°, and April 86½° and 46°. The nights throughout the winter are apt to be cool enough for open wood fires, and for blankets. Half the time an overcoat is not needed during the day, but it is never prudent for a stranger to be without one at hand.

"AIR. The atmosphere is singularly clear, tonic and dry. I have never seen it clearer anywhere in the world. It seems to have about the same bracing and exhilarating qualities as the air of the Great Sahara

WINTER HOME OF HON. WHITELAW REID.

in Northern Africa, or of the deserts about Mount Sinai, in Arabia Petræa. It is much drier than in the parts of Morocco, Algiers or Tunis usually visited, and drier than any part of the Valley of the Nile north of the

First Cataract. It seems to me about the same in quality as the air on the Nile between Assouan and Wady-Halfa, but somewhat cooler.

"ACTUAL HUMIDITY. — This is extremely slight, everywhere in Arizona, as compared with any eastern climate in the United States. The air is driest on the high mesas, remote from snowclad mountains or forests, and in the desert valleys, where no considerable irrigation has been begun. Wherever irrigation is carried on on a large scale, the percentage of humidity in the atmosphere must be somewhat increased, although to an eastern visitor it is scarcely perceptible. The same Government observations already cited show relative humidity, at Phœnix or Tucson, averaged for weeks, from morning and evening readings, as less than half the usual humidity on dry days in New York. General Greely, in a publication from the Weather Bureau, gave the normal weight of aqueous vapor in the Arizona air at from $1\frac{1}{2}$ to 4 grains per cubic foot.

"RAIN. — Showers, and, indeed, heavy rains, are liable to occur in every month of the year; but the actual number of rains seems to an eastern visitor strangely small. The average rainfall in southern Arizona, as shown by the Government observations, is but $8\frac{1}{2}$ inches per year.

"ALTITUDES. — It is a striking advantage offered by Arizona that, with the same general conditions as to temperature and dryness of air, the physician is able to select nearly any altitude he may desire. Thus, asthmatic sufferers can find almost the sea level at Yuma, or an altitude of only a thousand feet at Phœnix, or of only 2,400 at Tucson. Others, who find no objection to greater elevations, can choose between Prescott or Fort Whipple, 5,400; Flagstaff, 6,800; the Sulphur Spring Valley, or Fort Grant, 4,200; Fort Huachuca, 4,800, or Oracle, about 4,000.

"IS IT A PLACE FIT TO LIVE IN?—This depends on what one expects in a huge, sparsely settled territory of mountains and deserts. The man who looks for either the beauty or the seductive excitement of Monte Carlo will not find it. As little will he find the historic remains or the cosmopolitan attractions of Egypt; nor could he reasonably expect the amusements and luxuries of our own eastern cities. The people of Arizona are still chiefly busy in the pioneer work of subduing it to the residence and uses of civilized man. But it has two transcontinental lines of railway, with numerous feeders; it has fast mails, and rival telegraph lines, and is throbbing with the intense life of the splendid West.

"The two principal towns in the southern portion, chiefly sought for their climatic advantages, are Phœnix and Tucson. Each of them has ten thousand inhabitants or more. They have the electric light, telephones, trolley cars, plenty of hotels, banks, book stores, good schools, churches, an occasional theatrical performance, sometimes

VILLAGE OF JEROME AND CLIFF

a lecture or a circus, often a horse race, and, in the spring, a thoroughly curious and interesting 'fiesta.' For the rest, people must take their amusements with them. Good horses are abundant and cheap, and there are plenty of cowboys — the genuine article — to show what horses can do. The driving, for fifteen or twenty miles in almost any direction from Phœnix, is nearly always easy. The roads are apt to be dusty; but there is one well-sprinkled drive of six or eight miles; and since the winds are quite regular in their direction, it is rarely difficult to choose a route on which the dust will be largely carried away from you. The unbroken desert itself is often as easy to drive over as an eastern highway, and the whole valley is a paradise for bicyclers, or equestrians.

"CAN ONE LIVE COMFORTABLY?— That again depends on what you expect. You cannot have the luxuries of our New York houses out there, unless you build one; or the variety of our New York markets, unless you charter a refrigerator car. But there are hotels with almost as much frontage as the Waldorf; and, like everything else in the Territory excepting the mountains and the deserts, they are new. There are boarding houses of more kinds than one; and brick cottages of eight or ten rooms can occasionally be rented. Better than any of them, for the man with the energy and the pluck to take it, is a tent on the desert; and he who knows how to 'camp out' with comfort through September in the Adirondacks can camp out in Arizona through the winter.

"As to food, there is plenty, and it is good — if you can get it well cooked. The alfalfa fields of the Salt River Valley are the fattening ground for the great cattle ranges of the Territory. From there the markets of Los Angeles and even of Denver are largely supplied. Good beef, mutton and poultry are plenty and cheap. Quail, ducks and venison from the vicinity can also be had. Vegetables and fruits are abundant in their season, and sometimes the season is a long one. It is the one country I have lived in where strawberries ripen in the open air ten months in the year. I have had them on my table, fresh picked from the open garden at Christmas.

"IS IT A LAWLESS COUNTRY?—The man who goes to any considerable Arizona town with the ideas of the Southwest derived from novels, or from 'The Arizona Kicker,' will be greatly mystified. He will find as many churches as in towns of corresponding size in Pennsylvania or Ohio; and probably more schoolhouses. He will find plenty of liquor shops, too, and gambling houses, and dance houses, and yet he will see little disorder unless he hunts late at night for it, and he will be apt to find — as at Phœnix — a community of ten thousand people requiring in the daytime only one policeman, and hardly requiring him. During my winter there I did not see a single disturbance in the streets, or half a dozen drunken men, all told. Mining men and an occasional cowboy certainly had quarrels, sometimes, in the disorderly quarters at night; and there were stories of the use of the knife among Mexicans; but the visitor who went about his own business had as little trouble as on Broadway or Chestnut street. The Pima and Maricopa Indians, who are encountered

SALT RIVER VALLEY FOLIAGE.

everywhere, have been friendly with the whites for generations, and there isn't an Apache within some hundreds of miles.

"WHICH TOWN IS THE BEST?—Primarily that is a question for the physician, if there is a physician in the case—if not, try them all. If a mountain region, considerable altitude and a comparatively low temperature are desired, Prescott is in a picturesque region, near a great mining district, and has the social advantages of an army post, Whipple Barracks. Flagstaff is still higher, is in a region of dense pine forests, and is

within a hard day's journey of one of the
wonders of the world, the Colorado Cañon.
Oracle is a pretty mountain nook, embowered in splendid live oaks, like those of
California, and is also near an important
mining district. If lower altitude and a distinctly semi-tropical climate are desired, the
three places most likely to be considered are
Yuma, Tucson and Phœnix. The first is
near the sea level; is the warmest and
probably the driest of the three, has the
least population, and the smallest provision
for visitors. Tucson is the oldest town in
the Territory, and, after Santa Fe, perhaps
the oldest in the Southwest. Its adobe
houses give it a Mexican look, and are
thoroughly comfortable. Its newer houses
are of a handsome building stone, found in
the vicinity. The Territorial University is
here, and it was formerly the capital. Its
elevation being more than double that of
Phœnix, it is somewhat cooler, and as there
is next to no irrigation near it, the air is a
little drier. Phœnix is in the center of the
greatest irrigation in the Territory. The
country for miles around smiles with green
fields, covered with almost countless herds
of cattle, and it is everywhere shut in by
low mountains. It is the Territorial capital,
has the Government Indian School, the
Territorial Lunatic Asylum, and other institutions, and is the general focus for the
Territory. Like Tucson, it has its occasional
wind and sand storms — perhaps not quite
so often. At either place visitors who know

how to adapt themselves to circumstances can be entirely comfortable, and in each they will find an intelligent, orderly, enterprising and most hospitable community. They will find a country full of mines, full of rich agricultural lands, abounding in cattle and horses, in vineyards and orchards and the beginnings of very successful orange groves — a country, in fact, as full of promise for hardy and adventurous men now as California was in the fifties. Above all, if it has been their lot to search for health in far countries, they will revel in the luxury of being in their own land, among their own countrymen, within easy reach of their friends by telegraph or rail, and in a climate as good of its kind as any in the world. W. R."

* * * *

Apropos of the above, note the following clipping from *The Fourth Estate*, a New York City periodical, bearing date of December 3, 1896:

"Whitelaw Reid and Mrs. Reid are leisurely journeying to their winter quarters in Arizona, where they have found a climate softer and fairer for throat troubles than New York, and a charming scenery and pleasing people. They go in spite of the revived rumors that the journey occasions of Mr. Reid's broken health.

"The truth is, Mr. Reid's physician, who sent him off a year ago, after his return from Egypt, and an experience of a storm in the desert, told him he could live in New York this winter if he wanted to, as 'he had not a trace' of his asthma, bronchitis and tenderness of lungs that followed an attack of pneumonia. That was the precise medical report of the efficacy of Arizona air.

"Mrs. Reid, who is a great traveler and a good one, who has crossed our continent fifty times, and the Atlantic nearly as often, overruled her husband and his physician on the strong ground that it would be wise to lay in a store of health and strength by spending a second winter where the first had been so beneficial, especially as the balmy southern retreat and paradise was in our own country."

A Perfect Winter Climate.

Dr. W. Lawrence Woodruff, of Phœnix, contributes the following article on the winter climate of the Salt River Valley. He conservatively sets forth his reasons why

A COTTAGE IN PHŒNIX.

this is a favored place for the sufferer with consumption, asthma, rheumatism and nervous diseases:

"The winter climate of the Salt River Valley is as perfect as one could wish. To understand the possibility of the existence of such a climate in this country, consider for a moment the physical features which produce it.

"To the east, north and west, high mountain chains effectually shut out all cold winds. Lower mountain ranges to the south and southwest extend on both sides of the valley, to the Gulf of California, and on either side down the Gulf to its mouth.

"The Gulf of California, with its 53,000 square miles of surface almost within the tropics, and its open mouth 250 miles wide, draws into its funnel-shaped expanse wind and wave from the equatorial Pacific Ocean. This physical combination creates air currents which throughout the winter constantly sweep from the region of the equator over this vast tropical sea into our own land-locked valley, without being chilled by the cold air currents of the north.

"Thus is created in this southwest quarter of Arizona a region whose climate is fast becoming celebrated as the best the world can offer. To enjoy this perpetual sunshine, to bask in the balmy air, one has but a short journey, without the fears and danger of an ocean voyage. Italy, Egypt, the Orient, each is surpassed.

"Here is the natural home of the orange, lemon, lime, date, pomegranate, apricot, peach, and almost every other fruit and vegetable. Our plains and mountains have the same rugged, barren aspect as those of Palestine. We have the same genial sunshine as does California, without her chilly fogs; the same dry, bracing air without the altitude and blizzards of Colorado; the same warmth and luxuriant vegetation of Florida without her excessive moisture and malaria. We have all and more than all the advantages claimed by these places, with none of their drawbacks.

"*Multum in parvo* may be properly applied to the Salt River Valley. Here is found a combination of almost all desirable climatic qualities. There is rarely a severe frost. Nine out of ten days are clear and sunshiny. Habitual life out of doors is practicable and enjoyable, even for the invalid, and it is needless in this age to enlarge upon the beneficial influence of pure air and nature's sunshine.

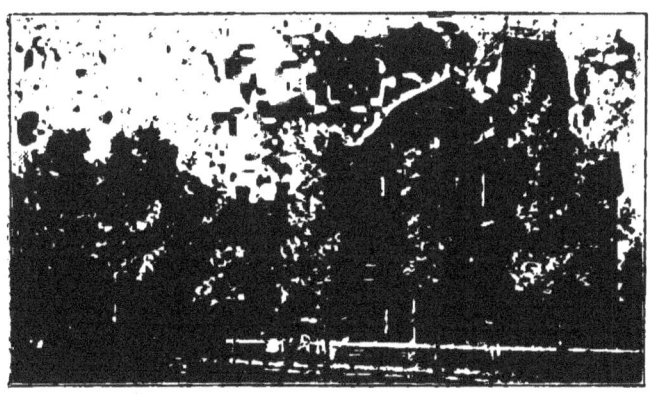

ON A SIDE STREET IN PHŒNIX.

"The constant dryness of the atmosphere, our even high temperature and few cloudy days will best be demonstrated by the following table:

COMPARATIVE DATA FOR PHOENIX, ARIZONA.
October, 1895, to May, 1896.

DATA.	October.	November.	December.	January.	February.	March.	April.
Mean actual temperature	72	57	49	54	56	62	64
Mean sensible temperature	59	49	41	44	44	48	48
Lowest temperature	48	34	23	30	28	34	38
Highest temperature	93	83	78	79	82	92	89
Mean relative humidity 5 A.M.	67	81	76	69	65	56	50
Mean relative humidity 5 P.M.	39	54	40	40	25	21	15
Percentage of sunshine	88	81	88	77	87	75	91
Monthly rainfall (inches)	0.80	0.89	0.09	0.46	0.05	0.39	0.05
Average hourly wind velocity	4.3	4.6	4.0	4.3	4.8	5.1	6.0

Annual precipitation, about seven inches.
Trace rainfall = Too small to measure.
100 = Continuous sunshine.
Station established August, 1895.

Arthur L. White,
Observer-in-Charge.

"Out-of-door life is enjoyable throughout the whole winter, except during the rainy season, which lasts but a week or so in either January or February. At other times throughout the whole of the year, one may spend every hour of the twenty-four out of doors. The invalid, not only with safety, but with benefit, may sleep in a properly constructed tent throughout the whole season, thus assuring pure air in abundance without the risk of drafts.

"I have said in another connection, 'Our one weak point is the difference between night and day temperatures. This difference is quite marked, but much more so measured by the dry bulb thermometer than by the wet bulb. The extreme dryness of the atmosphere makes the lower temperature less perceptible than in more moist climates, though there the extremes be considerably less. Owing to the dryness of the air the midday temperatures do not seem nearly so high as they actually are, neither do the lower temperatures at night produce the chill one would expect from looking at the reading of the dry bulb thermometer.

"'The actual discomfort from this wide range of temperature is but slight, and its dangers largely imaginary. Neither danger nor discomfort from this cause is equal to that in a moist climate with a range of temperature not more than one-third as great.

"'This difference is much less, and indeed exists but in a very small degree in the higher lands of the foothills and upper sides of the valley. The altitude at Phœnix is 1,100 feet, and in the foothills on the sides of the valley it will run from 300 to 500 feet higher.

"'The wind movement in the Salt River Valley is so slight as scarcely to be a factor. Our average annual wind movement is but two and one-half miles per hour. A wind of twenty-five miles per hour for any length of time is unknown. The gentlest of zephyrs

A PHŒNIX RESIDENCE.

usually prevail. On all sides there are barren mountains and desert. Nothing grows except by irrigation, and as the water is under the perfect control of man, there is no danger from decomposed vegetable matter. The atmosphere is so dry and pure that animal matter dries up instead of decaying. There being no marshes or stagnant pools, there is absolutely nothing but pure uncontaminated air to breathe.'

"That the pure, dry, warm air is invigorating and life-giving, and is indeed Nature's stimulant and tonic, I think is best proven by the following table of vital statistics:

VITAL STATISTICS of that part of the Salt River Valley north of the Salt River, west of the Verde River, and east of the Agua Fria River, covering a territory of 250 square miles, and including the City of Phoenix. The population on a conservative basis is put at 14,000; for 1895, at 15,000, and for 1896, at 16,000.

	1892	1893	1894	1895	1896
Total number of deaths	133	185	168	141	205
Transients	29	38	41	47	78
Accidental deaths	10	15	7	13	15
Among residents	94	132	120	81	112
Percentages, fractions 1%	¾	8-9	6-7	3-5	⅔
CLASSIFIED BY AGES.					
Deaths under 5 years of age	28	59	33	29	38
Deaths over 70 years of age	12	8	13	7	10
Deaths over 50 years of age	31	32	36	19	43
DURING THE SUMMER MONTHS —JUNE-SEPTEMBER.					
Total	41	75	54	58	75
Transients and accidentals	8	21	13	23	25
Residents, from natural cause	33	54	41	35	50
Percentages, fractions 1%	¼	2-5	⅓	¼	⅓
Under 5 years of age	6	28	13	14	19
Under 5, of bowel trouble	6	11	9	5	8
CAUSES OF DEATH.					
Stomach and bowel disease	10	30	21	14	15
Nervous and brain disease	17	8	4	8	6
Typhoid fever	2	4	4	2	4
Scarlet fever	1	3	0	0	0
Measles	0	4	0	0	0
Diphtheria	0	5	2	0	0
Heart disease	8	1	7	3	8
Disease respiratory organs	50	73	61	56	82
Old age	4	4	6	4	3
All other causes	40	56	58	54	87

NOTE.—Deaths designated as transients are only those of persons who have been here but a brief period prior to their decease, coming here as a last resort in the advanced stages of diseases of the respiratory organs, which accounts for the large number of deaths under this head. A large number of those claimed as residents ought properly to have been included in the transient class.

"In an article in the *Hahnemannian Monthly* (reprinted by the *Scientific American* and *The Sanitarian*) I said:

"'Now, as to diseased conditions: Asthmatics usually receive prompt relief and a permanent cure. The dry, warm air and low altitude agree with them perfectly. If there is a recurrence, it is during the rainy season and is usually but slight, to disappear again as soon as the usual dry atmospheric conditions prevail. This is equally so of aphonia, bronchitis and laryngitis, and in fact of all diseases of the respiratory organs. Tuberculosis, by the dry, hot air of summer, is checked in its development, and if the patient is not in the last stages, a continuous residence under these favorable conditions will greatly prolong life and often eventually bring about a cure. Let me say here, if the patient has entered the last stage of the disease, in the interest of humanity keep him at home. This cannot be emphasized too strongly. There he will have more comforts, and the radical change of climate with the long and tiresome journey necessary to reach here, only tends to materially hasten the end. During the winter months this class of patients, in common with all others, may reasonably expect to hold their own, and usually make substantial gains. It will readily be perceived by a careful perusal of this article that there is greater reason to expect beneficial results in all diseased conditions from a sojourn in this climate than in any other winter resort. While this is undoubtedly so, it is equally true that the hot, dry air of summer produces the best results. In heart diseases we find the cooler weather of winter the most beneficial. In some cases the reverse is true. The hotter and dryer it gets the more comfortable the patient becomes. This is especially so where the disease is complicated with diseased kidneys or rheumatic diathesis. Catarrhal conditions of head and throat are most relieved during the summer, especially the moist varieties. Diseases of the digestive tract, dyspepsia, chronic dysentery and diarrhœa, do exceedingly well here, and are usually promptly relieved. This is doubly true during the hot months. The summer conditions of high temperature and low humidity cause a determination of blood to the surface, and for months at a time maintaining it there, thereby entirely relieving all internal congestions. Kidney troubles are so prevalent I must not forget to mention that during the heated term the kidneys excrete less than one-half the normal quantity of urine. During this period of unrest the unloading of the effete material of the system is carried on by the sweat glands of the skin, and a healthy equilibrium is maintained. This continuous high temperature and very dry air keeps the blood at the surface, thereby making the sweat glands very active. Perspiration is constant and copious and by its instant evaporation keeps the surface cool and the bodily temperature at normal. These conditions are very advantageous to diseased kidneys, giving them a much-needed rest and an opportunity to recuperate. When to this is added a drinking water pure, wholesome and devoid of all alkali, it is easily understood why this valley is fast gaining an enviable reputation for the alleviation and cure of all forms of this disease. In rheumatic affections, while in winter patients are made very comfort-

THE FORD HOTEL, PHŒNIX.

able, it is in summer that the constant free perspiration, maintained for months without ceasing, entirely eliminates from the system all morbid material. In diseases of the nervous system, so prevalent in this age, this climate is a true panacea. This is especially so with persons suffering from insomnia and nervous prostration. Here, again, the best results are during the summer months. The universal verdict is, 'I have nowhere else slept as I do here.'

"This is the universal expression. The tired-out, starved nerves, overworked and overwrought, experience in this balmy air the perfect relaxation and rest they so long have been in need of. The dry, hot air of summer seems to quiet the nervous system — is soothing, restful. When to this a voracious appetite is added with perfect digestion, which is the only epidemic during this season, the results are understood without further elaboration. Finally, the perfect summer nights soothe and rest one's nerves as does nothing else in all the world."

Climate the Best Remedy.

While considering the climate of Phœnix and vicinity, another physician's views will be given. Dr. Harrison E. Straud, of Phœnix, says:

"It is a lamentable truth that a large per cent of the population of this world are not free to live where they choose, and where business and social interests demand; but are compelled to choose a climate in which they may enjoy health, or in many cases, where they can live at all; and especially is this true in the line of diseases of the respiratory organs.

"Periodically the world is startled with the discovery of some great antidote for that fearful scourge of humanity, consumption. The highest authorities in the world agree that at present we have no remedy or antidote that will destroy the bacillus of consumption without also destroying the patient — and in this dilemma attention is directed to the true and only remedy, which is dry air, combined with mildness and the least possible change of temperature between day and night and from day to day.

"Arizona is a large territory. Within its borders every possible altitude exists, from but twelve feet above sea-level, as at Yuma, to eternal snow in the mountains. It is difficult to appreciate that one condition exists in all these regions, regardless of altitude, namely, dryness. This fact is proved not only by Government observations, but by the testimony of the entire population.

"The dryness is perpetual; dead animals desiccate, so also does refuse matter. It is this dryness, with entire absence of dew and fog, that makes it desirable to sleep out of doors from May until October; and many sleep out the year round.

"The winter climate is delightful. It seldom frosts and never freezes. One bright sunshiny day follows another. Rains often occur between December and February; but entirely insufficient for agriculture. The days are warm and pleasant, the nights cool and exhilarating, the country green and pretty. Flowers bloom, and oranges as fine as any in the world ripen. Such a winter climate is pleasant and beneficial; but it is the heat of summer that is especially curative. This heat is peculiar; it is never oppressive except after one of the infrequent rains; at other times the great dryness makes the heat tolerable, or even comfortable. So little illness of any kind occurs in summer that 'physicians alone are sick.'

"It is strange, but perfectly true, that there has never been, to my knowledge, a case of infantile diarrhœa during the hot period. That terrible disease, cholera infantum, is positively unknown to us. Again, the longer one lives here the more he prefers the summers; and as a matter of health they are incomparable.

"We have no tornadoes, cyclones, or sunstrokes.

"A very important point is the food supply, not only for invalids but for others, especially regarding meats. The pale clerk who eats meat twice daily will outwear and outlast the burly laborer whose size is grown on potatoes, corn and other starches. I can truthfully say I have never seen better meats, in Europe or America, than are daily sold in Phœnix. The gardeners supply green vegetables every day of the year, and fruits are plentiful and good.

"Speaking of Phœnix, if the question were asked: What is such a climate and environment especially recommended for? I should say, first the commencement of consumption, chronic bronchitis, asthma and rheumatism. These especially in my experience find relief and cure. If the question were asked: What disease is the climate of Arizona not adapted for? I should be obliged to say, I don't know. The altitude is not high enough to cause nervousness or hemorrhage in those of a hemorrhagic

A VIEW ALONG WASHINGTON STREET, PHŒNIX.

tendency, neither is it too high for most cases of heart disease."

The Editor's Point of View.

The editor of a daily newspaper can usually see a point quickly and state it tersely. Below is a comparison made by the editor of the Phœnix *Gazette*, which speaks for itself:

"On February 9, the steamer Germanic came into New York harbor in a temperature of nine degrees below zero, a blizzard blowing that had delayed the good ship many days. The harbor was full of floating ice, the ferry boats were stopped, not a train was running into the great city, and business at a standstill.

"Down the coast the temperature rose somewhat, but in Florida the atmosphere still was cold enough to freeze the oranges on the trees, and the trees themselves, damaging the industry in the alligator State to the extent of fully $10,000,000. In one week, citrus fruit-growing in the southeast was practically wiped out. About the same condition of affairs prevailed along the Gulf States, working damage to the extent of millions to agriculture, as well as immense loss to the shipping and mercantile interests.

"Working westward along the southern boundary line of the United States, not till Arizona is reached did the average temperature on that day rise to above thirty-two degrees, the freezing point of water. The line that marked the freezing temperature coursed up through the eastern and then through the northern part of Arizona, included a portion of the southern division of California, was lost in the waves of the Pacific, again touching the land only at a point on the northern coast of Washington, where the land most appreciably feels the warming effect of the wash of the mighty Japan current.

"On that day the temperature in Phœnix was but little below the normal for the season, and the palms and vegetables were untouched by frost. The citrus groves were unhurt, and only an occasional cloud marred the blue expanse of the heavens.

"Little wonder it is that from the stormy East there should turn so many toward the land where man is not, as is the squirrel, compelled to earn his livelihood with an eye ever to the stormy days of winter; where nature is ever helpful and never unkind."

Sunnier than the Riviera.

Lest the reader should imagine that none but invalids are made welcome in Arizona, the following letter, written by Granville Malcolm, of Denver, throws a side light on the subject from the standpoint of a pleasure-seeking tourist. Mr. Malcolm says:

"Much has been said about the climate and healthfulness of the Salt River Valley.

But I have seen no opinion as to the advantages of this as compared with other favored resorts.

"Having spent several winters in Colorado, three in California, one in Florida and Cuba, one in Texas, parts of three winters in Phœnix, part of one winter in Thomasville, Georgia, and New Orleans, and one winter on the Riviera in the south of France, and Italy, my judgment without bias is strongly in favor of this valley as a winter resort, having a climate surpassing each of those named.

"The reasons for my conclusion regarding the Salt River Valley might, if given in full, take too much space. But the main reasons shortly stated, and which appeal to even a casual observer as well founded, are the favorable conditions of temperature and humidity existing here as in none of the other localities named. For instance, on the Riviera there are not half so many sunny days; there is more humidity, and consequently the air has more 'chill' in it than there is here. It rains a great deal on the Mediterranean coast, and the dampness is sometimes very trying. If this valley was supplied as the Riviera is with delightful hotels and pensions for sojourners, the tide of seekers for a winter climate par excellence would very soon turn this way.* The mildness of the temperature (evidenced by the tender semitropical products of this valley that thrive

---
* NOTE.— The accommodations are now much better than when this was written.

the winter through); the dryness of the atmosphere, that precludes the feeling of chill one feels near the coast; the almost unintermitting sunny days, the favorable altitude—all these conditions conduce to the verdict in favor of this as a resort superior to all the others. And when these advantages are appreciated, the valley will be filled with those seeking homes here, or seeking immunity from the severity of northern winters.

"Each time I return to Phœnix but raises my estimate of this charming valley as a winter resort, and I shall do what I may to sound its praises."

Phoenix in a Nutshell.

As an answer to some questions asked by those who wish to be particularly informed respecting the social life, accommodations, etc., at Phœnix, the Chamber of Commerce of that city furnishes considerable data which is condensed for reproduction below:

The Salt River Valley is over sixty miles long and averages twenty miles in width. To the eye it presents a perfectly level appearance, though there is a gradual slope South and West. In the center of this valley, at an altitude of 1,080 feet, is Phœnix. The city is surrounded by mountain ranges, the nearest being six miles away. The location is a pleasant one from a scenic point of view, and likewise desirable as a business proposition.

39

Phœnix has 16,000 inhabitants; they are a progressive class of Americans. Residences, business houses and public buildings are substantial and attractive. A large transient population, principally composed of tourists and healthseekers, gives the city a busy appearance and adds much to its material resources.

Under the head of improvements — an extensive waterworks system is in operation; illumination is supplied by gas and electricity; a telephone system extends to neighboring towns and ranches; there are ten miles of electric street railway, also a well-equipped fire department with electric fire alarm system.

Church organizations are maintained by the following religious bodies: Presbyterian, Baptist, Episcopal, Christian, Methodist Episcopal, Methodist Episcopal South, Roman Catholic and Seventh Day Adventist.

There are four lodges of the several degrees of Masonry, four of Odd Fellows, and one each of Workmen, Knights of Pythias, Select Knights, G. A. R., Chosen Friends, Good Templars, Elks and Red Men; the W. C. T. U. conducts a public library and reading room; the press is well represented by influential daily and weekly newspapers.

The Maricopa Club has a large membership and handsome quarters, extending the customary courtesies to visiting strangers.

Phœnix has three large school buildings, with a $30,000 high school under construction; 1,500 pupils were enrolled in 1897. A number of handsome State and county buildings are already erected, including the Territorial insane asylum, costing $100,000, while the site for the capitol has been selected and a large sum already expended in beautifying the grounds.

The various professions and all kinds of retail business are well represented. The

COMMERCIAL HOTEL, PHŒNIX.

city also possesses two wholesale grocery houses, two ice factories, three planing mills; five lumber yards, three foundries, one creamery, one onyx factory, two large roller process flouring mills, four banks, five hotels and three public halls.

The above statistics will serve to show what Phœnix is, and to dispel any erroneous impressions that may prevail concerning the undesirability of life in the Far West.

The cost of living is about the same here as in the East.

There are numerous places for taking

care of invalids; a sanatorium which can accommodate a considerable number of patients; a Sisters' hospital, several hotels, lodging houses, restaurants and furnished rooms—while out in the country there are good accommodations among the ranches. Phœnix has fifty physicians, including some specialists in lung and throat diseases. Competent nurses may be readily obtained.

Driving, horseback riding and bicycling are popular modes of recreation here.

Phœnix is quite a pretty place, its beauty all the more noticeable when contrasted with the arid portions of Arizona. Shade trees abound and in almost every yard fruit and flowers may be found in season—olives, pomegranates, figs, great hedges of geraniums and acres of lilies.

Hotel and other Accommodations.

A general statement that one may find hotel or other accommodations at Phœnix on a par with larger eastern cities hardly meets the requirements of tourists wholly unacquainted with the locality. Hence a list has been carefully compiled which aims to give detailed information on that point. Prices quoted are subject to change. The list is as follows:

Adams Hotel.— One hundred and fifty rooms, four blocks from depot; new and first-class in every particular; rate of board averages from $3 to $6 per day; hotel not open during summer except for rooms.

Commercial Hotel.—One hundred rooms, three blocks from depot; first-class accommodations for 150 people; European plan; rooms $3 to $9 per week, winter or summer; first-class dining room in connection; board $5 to $7 per week.

Ford Hotel:—Sixty rooms, accommodations for 75 to 100 people; board and room $2 to $4 per day.

Lemon Hotel.—Fifty rooms, accommodations for 75 people; $3 to $6 per week for rooms; $4.50 to $7 for board.

Sixth Avenue Hotel.—Forty rooms; room and board, $35 per month.

Mills House.—Thirty rooms; capacity 50 guests; rooms per night, 50 cents; summer rates, $8, $10 and $12 per month; winter rates, $10, $12 and $15 per month; board, single meals, 25 cents, $4.50 per week. Street cars at door every ten minutes.

The Alhambra.—Twenty rooms; room and board, $30 to $50 per month.

The Westminster.—Twenty rooms; 25 and 50 cents per night.

The Perkins.—Twenty rooms; no board.

The St. Lawrence.—Twenty rooms; $3 to $6 per week; no board.

Hardwick Hotel.—Thirty rooms; board and room from $5 to $7 per week.

Capitol Hotel.—Twenty rooms; no board; $3.50 to $7 per week.

The Alamo.—Twenty-five rooms; $3.50 to $6 per week.

College Place Lodging House.—Six blocks from depot; 50 rooms; $1 per week and up. C. D. Ward, proprietor.

Wharton Rooming House.—No. 38 North Center street, twenty-one rooms; accommodations for 40 guests; summer rates, 25 to 50 cents per night, $1.50 to $2.50 per week and $6 to $10 per month; winter rate, 50 to 75 cents per night, $2 to $3.75 per week, $8 to $15 per month.

Rooms with bath are to be found in all parts of the city, and requests for accommodations should be placed as early as possible in order to secure good accommodations. There are also a great number of restaurants where first-class board can be had on the European plan from 15 cents up, and on the American plan from 25 cents up.

In this connection the Adams House deserves some special mention. While there are other very good hotels in Phœnix, the Adams has the distinction of being the largest. The edifice cost $200,000; it is built of pressed brick with brown stone trimmings; is four stories in height; has wide verandas on every floor, a passenger elevator and spacious parlors, halls and dining room. The culinary department is a noteworthy feature.

The erection of the Adams House and the improvements in other hotels have removed one drawback to tourist travel, namely, lack of sufficient accommodations. Phœnix is amply supplied in that respect.

As Seen by a Journalist.

Under the title of "Ten Days in Arizona," Mr. Julian I. Williams contributes a readable article to the *Southwest Illustrated Magazine*, touching upon phases of Phœnix life not heretofore mentioned. A few extracts follow:

"No one who is a lover of nature can fail to appreciate the grandeur and wonderful beauty of the Salt River Valley with its varied mountain forms and the magnificent views from the valleys, which, until a few years ago, had been for centuries as a sealed book. Lately this country has had a wonderful attraction for the outside world, and a visit to the territorial seat of government, Phœnix, will serve to reveal the reasons for this.

"I had occasion to use a carriage one lovely morning, and asked the driver to take me to see the better class of residence streets. The driver replied that 'There wasn't none but what was good.' The man, his honest face glowing with pride, touched his hat, mounted his box, adjusted his silver-buttoned livery, and sent his prancing steeds off on their welcome errand. A drive around the city revealed several facts, one of which is, that the city keeps the pavements in better condition than do older and wealthier cities. We whirled rapidly down Washington street, and as we were nearing the end of the street the impatient horses were pulled into a walk, while the spacious Territorial grounds were pointed out. It is the site for the new capitol, and a most beautiful one at that. A gardener is kept constantly employed, and as a result the grounds are in the best possible condition. The magnificent forest trees shading the half dozen or more acres of ground, the elaborate arrangement of flowers, date palm and fan palm, give an air of repose unknown to many spots in this busy, bustling town. Back from Washington street we went fully three miles, and the favorable impression first received, the succeeding blocks fully sustain. Washington street is the main business thoroughfare. It is three miles in length and is lined on either side with business houses, which will compare favorably with any in the larger cities of the Southwest. Many of the streets are fringed with trees, which to the stranger seems a very pleasant feature. Ditches on either side flow with clear water, occasionally diverted to irrigate the yards and grounds of homes, where flowers and fruits are seen in abundance."

The Arizona Summer Climate.

One might not think of Arizona as a pleasant place to spend the summer in. The popular impression is that intense heat renders existence unbearable. To show that the public errs, the article below is reprinted from the *Medical Century*, where it appeared September 15, 1896. Dr. W. Lawrence Woodruff is the author:

"The month of June, 1896, will be celebrated as having the highest range of temperature and for the greatest number of consecutive days ever known in the Salt River Valley, if not in the United States.

"The following table shows the actual heat for the first half of the month as marked by the reading of the dry bulb thermometer, the so-called sensible temperature, as indicated by the wet bulb, and the relative humidity or percentage of saturation, according to the observations of the United States Weather Bureau, at Phœnix, Arizona:

Date.	Actual.	Sensible.	Rel. Hum'ty
	Degrees.	Degrees.	Per cent.
1	97.1	65.6	16
2	95.9	64.0	12
3	94.0	61.0	11
4	91.0	60.8	14
5	93.8	61.0	12
6	94.8	63.8	15
7	97.0	64.5	31
8	100.8	65.4	12
9	104.8	64.8	8
10	107.0	67.0	9
11	109.0	67.8	7
12	109.5	68.8	10
13	114.8	72.0	13
14	114.5	73.0	11
15	114.0	71.5	10

"From June 9 to 18, inclusive, was the longest continuous period of extremely hot weather within the memory of the oldest inhabitant. From the 13th to the 17th, the best accredited thermometers (set nearer the ground than the Government instrument), registered from 3 to 5 degrees higher, and indicated from 118 to 120 degrees Fahrenheit. It will be noted that the difference between the actual and sensible temperatures (indicated by the readings of the dry and wet bulb respectively) was from 30 to 43 degrees, depending principally upon the percentage of humidity. On only seven days did the relative humidity go above thirteen per cent. This is a fair index of the dryness of the summer air in the Salt River Valley.

"With this record of intense heat, extending over one-third of the month, should be coupled that of the wonderful exemption from disease during the same period. Nowhere else in the known world were the inhabitants so healthy as in Phœnix and its vicinity. There was practically no acute sickness.

"The following table of deaths for June, 1896, in that portion of the Salt River Valley north of the Salt River, west of the Rio Verde, and east of the Agua Fria, containing a population of 16,000, and including the city of Phœnix, is a fair index of our ordinary summer healthfulness:

Cause of Death.	No. Cases	Age.	Remarks
Puerperal fever..	1	27	
Typhoid pneumonia..........	2	27-8	
Bowel disease................	1	2	
Typhoid fever and chronic alcoholism............... ...	1	79	
Chronic alcoholism and heat prostration..................	1	64	Tramp.
Old age........................	2	85-86	
Brain fever...................	1	24	
Consumption	4	.....	All transients.

"During the months of May, June, July, August and September, 1895, there was but one death each month from bowel trouble among children in the territory named.

"During the five summer months of the past four years the total death rate was as follows:

    1892, one-fourth of one per cent.
    1893, two-fifths of one per cent.
    1894, one-third of one per cent.
    1895, one-fourth of one per cent.

An average of two and 85-100 in 1,000 inhabitants. This is the season, in all other parts of the world, of greatest fatality from gastro-enteric diseases.

"Were it possible, the world ought to know, not only that the Salt River Valley during the summer time is the healthiest spot on earth, but that the healthy individual and the healthseeker can live in the Salt River Valley during the summer. Our hot, dry air is stimulating, and not in the least debilitating. We usually find (when there is sufficient vitality left to expect any benefit at all) a gain in weight and strength so long as the hot weather lasts. A summer spent here with its unloading of poisonous, effete, broken-down tissues, prepares an invalid to get the greatest benefit from our genial winters."

More About Summer Heat.

The following extracts are from an article by Maj. H. F. Robinson on the subject, "The Real Temperature in the Salt River Valley":

"A very erroneous impression has gone abroad concerning the 'terrible heat' experienced in the Salt River Valley in the summer time. This is based, in the main, upon the records of the metallic or ordinary thermometer. There is a vast difference, however, between the sensible temperature and that indicated by the thermometer we are accustomed to base our ideas of degrees of heat and cold upon, in the dry atmosphere of so-called 'arid America.'

"The records kept in the Southwest, of the temperature in the summer time, show extremes of heat; but it is a fact well known to the inhabitants, and now beginning to be

OSTRICH FARM.

understood in the East, that the sensation of heat as experienced by animal life is not accurately measured by the ordinary thermometer.

"The ordinary thermometer gives the temperature of the air only, and takes no notice of the other factors present.

"The human organism, when perspiring freely, evaporates the moisture from its surface and thus lowers its temperature. The meteorological instrument that registers the temperature of evaporation and thus in a great measure the actual heat felt by the human body is the wet bulb thermometer. This is an ordinary thermometer, the bulb of which is covered with cotton which takes up moisture from a cup of water by capillary attraction. The drier the air, the more rapid is the evaporation and consequent coolness; the drier the atmosphere the lower the sensible temperature when compared with the air temperature; the damper the air the higher will be the sensible temperature. This will be better understood by the statement that when the air is moist to saturation (that is, holding all the moisture it can without precipitation) the readings of the ordinary and wet bulb thermometer are the same, and the sensible temperature and the air are equal. In the East, where the air is always charged with more or less moisture, the difference is not great; but in the West and Southwest, particularly in Arizona, humidity is almost absent. On account of the extreme dryness in the summer time the sensible temperature is often 20 to 30 degrees lower than the air temperature, and sometimes even more.

"It has always been a mystery how, with our apparent great heat, there has been an entire absence of sunstroke, and how it was possible to work the entire day in the hay field, as is done, and not suffer for it. The extreme range between the actual and apparent temperature, however, fully explains this; while the very limited range in the East explains why prostrations from the heat occur with much lower temperature as indicated by the ordinary thermometer.

"In 1896, the hottest weather experienced in Phœnix was in June, and the following records were kept by the Weather Bureau; readings taken at 5 P. M. local time (First column indicating dry bulb reading; second, wet bulb; third, difference):

June 14—dry 111.2°; wet 73.5°;—37.7°
15—dry 113.0°; wet 72.0°;—41.0°.
16—dry 113.7°; wet 72.5°;—41.2°.
17—dry 111.1°; wet 69.5°;—41.6°.
18—dry 107.2°; wet 69.8°;—47.4°.

"Along the Atlantic coast and along the great lakes the mean difference between the wet and dry bulb thermometers is not far from 5 degrees, so that on the hot days noted above, the heat felt in Phœnix was no greater than it would have been in Chicago or New York had the range of the ordinary thermometer been from 75 to 79 degrees."

Major Robinson's position with regard to Arizona's summer climate is sustained by Captain Glassford, of the United States Army Signal Service, at Denver, Colorado. He confidently asserts that Phœnix enjoys a climate equally as agreeable in summer as that of San Antonio, Texas, or Augusta, Georgia.

Making a Living in Arizona.

There are some invalids who, after paying the expenses of getting to Arizona, would have but little money left for current expenses and doctors' bills. To such it is of importance to know whether they can, after reaching destination, earn enough in some light occupation to keep their purses moderately full. There is not space here for a detailed account of all the cash-producing occupations. A few are briefly mentioned.

There is money in barley. Witness the case of a farmer who planted 500 acres to barley and harvested $9,000 worth of grain in one season, thereby paying the cost of the land and realizing a profit of 50 per cent. Over 30,000 acres in the Salt River Valley are sown in barley, with an average yield of forty bushels to the acre.

There is money in alfalfa. In 1893 a forty-acre tract returned its owner a gross income of $1,600. Another man made $5,900 net on 160 acres in three years, by feeding his alfalfa to cattle. Not less than 50,000 acres are planted in alfalfa, capable of producing 350,000 tons of hay yearly. Three to five crops may be harvested in a single year, yielding two to five tons of cured hay per acre at each cutting.

There is money in bee culture, for here the bees feed on mesquite and alfalfa blooms, producing a honey noted for its mildness and fine quality. The honey from this valley is shipped to all parts of the United States.

There is money in live stock. Under these mild skies, and with an abundance of nutritious food, young animals mature quickly and at small expense for rearing. Horses, mules, cattle, sheep and hogs are a source of profit. Contagious diseases do not exist. Shipments of cattle from Arizona

A COUNTRY ROAD IN THE SALT RIVER VALLEY.

amounted to nearly $3,000,000 in 1896. Blooded stock is being rapidly introduced. Sheep ranching is also a lucrative industry, and breeders of hogs have found Arizona's protected valleys well adapted thereto.

There is money in poultry. Chickens sell at $4 to $6 per dozen; turkeys, 10 cents per pound live weight, while eggs average 25 cents a dozen. Hatching can be carried on all the year without shelter.

There is money in mining. Gold, both in placers and rock formations, is found in paying quantities between Phœnix and Prescott. One mine reports $11,000,000 worth already taken out. Others have done almost as well. Copper mining is a great and lucrative industry. Silver, coal, iron, tin, marble, onyx and lead are also found. The mines of Arizona have produced for the twenty years ending June 30, 1896, gold, silver and copper, aggregating $127,000,000. The annual return from dry placer mining alone is about $600,000. Forty million acres of Arizona land is mineral bearing.

There is money in sugar beets, onions, sweet potatoes, and all kinds of vegetables. But the dollars come quickest and easiest from the raising of small fruits. It is the orchard and vineyard that particularly invite the young man with a fortune to make, or the old man with a fortune to enjoy.

Not even the most favored sections of California are better adapted to the profitable raising of oranges. Trees grow thriftily and are not affected by disease. The fruit matures evenly, being bright, clean and highly colored. In juiciness, richness of flavor and marketable qualities, Arizona oranges and lemons have no superior. This is admitted by California and Florida experts. As to profits, it is nothing unusual to receive $400 per acre net from a five-year-old orchard. One company alone now has 1,500 acres set to oranges, lemons and limes.

Other fruits that find a home here are: Apples, pomegranates, limes, peaches, straw-

A BUSINESS STREET IN PHŒNIX.

berries, almonds, plums, figs, quinces, nectarines, lemons, cherries, bananas, pears, olives and apricots. Peaches come into bearing early and do well. Apricots yield 75 pounds per tree the third year and at maturity over 400 pounds; they are dried and also shipped in refrigerator cars. Olives thrive well and are free from scale.

The grape and wine industry is assuming important proportions. As a sherry wine district, experienced viticulturists pronounce the Salt River Valley to be without an American rival. Seedless raisin grapes are successfully produced, and bring good prices when packed for shipment to outside markets.

Getting fruit to market does not eat up all the profit. Besides the innumerable mining camps and stock ranches which this valley supplies, the foreign market (as far east as Chicago) is reached with a considerable advantage in distance, time, and earliness of ripening, as compared with competing fruits from some other localities. Navel oranges are ready here for market by November 10. Farm labor ranges from $20 to $30 per month and board; domestics get $15 to $40 per month; common laborers earn $1.50 to $1.75 per day and skilled labor $2.50 to $6; clerks receive $25 to $100 per month.

The Salt River horticulturist is placed in a position where intelligent industry will enable him to secure results equal to those of any fruit district in the world.

Prescott and Vicinity.

The line of the Santa Fe, Prescott & Phœnix Railway extends from Ash Fork (on the Santa Fe Pacific division of the Santa Fe Route) to Phœnix, a distance of 197 miles. It runs through a district where exceedingly rich silver, copper and gold mines are operated. The greater part of this road traverses either high table lands or penetrates a mountainous region. The altitude invites a different class of invalids from those who seek the Salt River country, and especially in summer, tourists find many opportunities for rest and recreation at Prescott, Jerome, Castle Creek Hot Springs and adjacent points.

Dr. J. Miller, of Prescott, furnishes an article regarding that city:

"Prescott, the county seat of Yavapai County, is most beautifully located in a large basin surrounded on all sides by mountains, the highest of which rises to a height of 10,000 feet above sea level. Granite Creek, a small stream, flows through the city from south to north. A beautiful pine forest stretches for miles to the south and west.

"The population is 3,500, and in nationality, unlike many towns in the Southwest, is strictly American, there being less than one dozen Mexican families in the place. There are five Protestant churches — Episcopal, Methodist Episcopal, South Methodist, Congregational and Baptist — and one Catholic church. There are two public school buildings, both brick structures. The fraternal orders are well represented.

"Owing to the mountainous condition of the surrounding country there are no good drives in the immediate vicinity (in the common acceptation of that term), but good, safe mountain roads, leading from the city in every direction, afford beautiful views of ever-varying scenery which one cannot fail to enjoy. Deer are still plentiful in the mountains, while quail, duck and wild geese are abundant in the adjacent mountains and valleys.

"The altitude is about 5,400 feet above sea level, and sunshine predominates throughout the year. The temperature for the year 1896 was as follows :

January, February, March: average maximum, 54.48 ; average minimum, 23.95 ; total average, 39.52.

April, May, June : average maximum, 77.92 ; average minimum, 40.2 ; total average, 59.6.

July, August, September: average maximum, 86.75 ; average minimum, 53.87 ; total average, 70.28.

October, November, December : average maximum, 59.46 ; average minimum, 27.38 ; total average, 43.46.

"On account of the dryness and rarity of the atmosphere the effect of both heat and cold is not so great at the same temperature as in a humid atmosphere, no greater discomfort being felt with a temperature of 100 degrees above zero than at 65 or 70 degrees in the Eastern States, and sunstrokes have never occurred here.

"While the climate, owing to the dry atmosphere, is specially beneficial to persons suffering with pulmonary or throat diseases, the facilities for caring for invalids are limited, there being no sanatorium established yet, though one is talked of. A limited number of persons can always find accommodations in hotels, boarding houses and with private families at a cost of $1.50 to $3 per day, or for $30 and upward per month. See list on page 61.

"There are ten regular practicing physicians located in Prescott, while two army surgeons, stationed at Whipple Barracks, one mile from Prescott, have a limited practice in the city. None of them makes a specialty of treating lung and throat diseases. There are also several experienced nurses in the place.

"Whipple Barracks, a military post, is located one mile north of town. The headquarters of the 11th Infantry, United States Army, is located there, and the band and four companies of troops are stationed at this garrison.

"Here malaria with its many complications is unknown. The dreaded summer diseases of childhood are conspicuous by

WHERE BEEF IS MADE.

their absence, and instead rosy cheeks and smiling faces gladden the mother's heart.

"For those suffering from pulmonary diseases, such as asthma, bronchitis, consumption and diseases of the throat, this is the ideal place of residence. The rigid blasts from the north or parching heat from the south never find their way here. Unlike some other valleys of the Southwest, this one is free from sand storms. Here will be found complete immunity from fleas, mosquitoes and other pestiferous insects. The air being warm and dry, the much-dreaded fogs of the coast and low lands are entirely avoided."

The following information about accommodations at hotels, etc., in Prescott will be of interest:

Burke Hotel.— Burke & Hickey, proprietors; three blocks from depot; sixty rooms; $2.50 to $3 per day; $14 to $20 per week; $40 to $60 per month; American plan.

Winsor Hotel.— Brown & Kastner; three blocks from depot; thirty rooms; room, $1 per day; dining room conducted on European plan.

Sherman House.— George Schuerman; four blocks from depot; thirty-five rooms; room and board, $1.50 per day; $9 per week; $35 per month.

The Brinkmeyer.— H. Brinkmeyer; two blocks from depot; forty rooms; $1 to $1.50 per day; $7 to $8 per week; $26 to $30 per month.

Prescott House.— H. J. Iftiger; one block from depot; thirty rooms; $1 to $1.50 per day; $6 to $8 per week; $25 to $30 per month.

Johnson House.— Miss C. Johnson; two blocks from depot; thirty rooms; $1.50 to $2 per day; $7 to $10 per week.

Congress House.— W. Richardson; two blocks from depot; eighteen rooms; no board; rooms, 50 cents to $1 per day; $3.50 to $5 per week.

Mrs. Gould.— Three blocks from depot; ten rooms; board and room, $30 to $40 per month.

Prescott Further Described.

Dr. E. W. Dutcher, of Prescott, writes in detail respecting the advantages of its climate. He says, in part:

"January and February are clear and cool, with frosty mornings and beautiful warm days. The air is dry and invigorating. Occasionally it rains for a few hours, while the mountain tops will be white with snow. March in Arizona is like March almost the world over, only in a milder form. But March is the only disagreeable month. April, May and June are clear and free from storms. In July and August comes the rainy season, namely, a shower lasting an hour or two once or twice every week. September, October, November and December are beautiful months in Prescott. And I believe the air is clearer in this part of Arizona during those months than anywhere else on earth.

"Altitude is one factor if not the chief one in curing consumption. I favor an altitude of not less than three thousand feet above sea level.

"The person wishing to become an athlete will work faithfully to develop muscle, and muscle cannot be developed without exercise.

"How can the patient suffering with pulmonary tuberculosis expect to restore the lost power in the lungs in Florida, Califor-

nia, or any other part of the universe while living at sea level; breathing an atmosphere loaded with fog and mist? They can breathe and breathe easily if a part of one lung remains. But how about the cure? The disease remains. The lungs waste away and the invalid who has traveled thousands of miles in search of climate, dies. In this climate, at this altitude, the invalid *must* exercise the lungs. They will say, it is hard to breathe; but that lasts for a few days only. They will expectorate more freely than ever. The air cells that have been shut, clogged and useless for months are being cleared. The pure, bracing life-giving air is again penetrating them, the fourteen hundred square feet of surface in the lungs is healed and restored to health by Nature's own remedy, namely, a constant bath of pure, cool and dry air that has been sifted through boughs of pine, juniper and spruce.

Castle Creek Hot Springs.

There are hot springs and hot springs. Some possess great medicinal value. Others are merely flowing water, heated above normal temperature, having no curative properties except such as pertain to water generally. The springs that really cure stubborn chronic diseases are not very numerous, but distance does not deter the invalid if it is found that the waters of any particular hot springs are truly beneficial.

In central Arizona, twenty-four miles from the railroad, is a new candidate for public favor, Castle Creek Hot Springs. Those who have summered and wintered there, or who have spent only a few days on the spot, have been so well pleased with results obtained that the owners of the resort are preparing to widely advertise it, and as fast as possible suitable accommodations will be prepared. For the present, the accommodations, while good, are rather limited; but this deficiency will soon be remedied. Extensive improvements are being made, and it is proposed to open a new hotel, wholly modern in its equipment.

The basis of this resort is, of course, the springs themselves, which bubble up from subterranean sources boiling hot. The waters are remarkably free from organic matter and have proven successful in curing maladies of the blood as well as in generally toning up the system.

And there is something to see, too. The surrounding mountains attain an altitude of several thousand feet. They present a magnificent appearance, and under the magic of Arizona's wondrous sunshine assume strange and brilliant color effects.

It is easy to get there. A fine mountain road has been constructed from Hot Springs Junction (on S. F. P. & P. R'y) to the Springs, a distance of twenty-five miles in an easterly direction. A comfortable stage makes daily trips along this route, the regular schedule being less than four hours in each direction. After a long railroad journey, this stage ride among the pine-clad mountains is a most delightful experience.

The Mining Industry.

It is not out of place, in a pamphlet describing health resorts, to mention the wonderful interest being manifested in the mining industry all along the line of the Santa Fe, Prescott & Phœnix Railway.

The superiority of Phœnix as a health resort attracts many persons of means who are not seeking rest or a change on their own account, but who accompany invalid friends or relatives. A splendid opportunity is offered such persons to investigate the different repositories of treasure adjacent to Phœnix and the Santa Fe, Prescott & Phœnix Railway; and in a number of cases they "strike it rich," as the miners say. The large amount of prospecting and development work constantly going on also furnishes employment for considerable skilled and unskilled labor.

Wickenburg, the Castle Creek Hot Springs country, Congress, Stanton, Harqua Hala, Santa Maria, Peoples' Valley, Big Bug, Kirkland — in fact, all the country surrounding Congress, Prescott and Jerome, is a storehouse of mineral wealth, only awaiting the intelligent direction of brain and capital to enrich the careful investor.

The United Verde Copper Company, of Jerome; Congress Gold Company, of Congress; McCabe Mining Company, and Providence Mining Company, of Prescott, and the Crowned King Mining Company, of Crowned King, Arizona, are notable examples (among many) of mines paying gratifying dividends to their stockholders. With the building of contemplated branch railroad lines to several rich ·camps adjacent to

SCENE ON THE CANAL.

the Santa Fe, Prescott & Phœnix Railway, there will be increased activity in mining and many more desirable opportunities for the investment of large amounts of capital.

At Congress, about four miles from Congress Junction, on the Santa Fe, Prescott & Phœnix Railway, and in plain view of the railroad station, is located one of the model mining towns of Arizona. Here the Congress Gold Company operates a gold mine on an extensive scale. Forty stamps are working night and day; improvements under way will increase the capacity to one hundred stamps. The ore, a quartz sulphuret, is rich in gold, and what is not secured in the concentrates is saved from the tailings by the cyanide process.

Three hundred men are employed, and as the town is under the direct supervision of the company, good order is maintained. Churches, schoolhouses and halls have been erected for the education and amusement of the inhabitants. Limited hotel accommodations can be secured, and, as the climate during the winter differs but little from that of the Salt River Valley, seekers after health will find a pleasant diversion in making a trip to Congress, surrounded as it is by picturesque mountain scenery.

Jerome Mining Camp.

Jerome, a typical mining camp, with a population of 2,000; 500 of whom are Mexicans, is situated in the Black Hills, Yavapai County, at an elevation of 6,000 feet above the sea level and 3,600 feet above the Verde Valley. The United Verde Copper Mines are located here. These mines are acknowledged to be the richest in gold, silver and copper in the world.

At the foot of the mountains on which Jerome is built lies the Verde Valley, watered by the never-failing Verde River, from which water is taken to irrigate thousands of acres of land now under cultivation, and from which water enough could be taken to irrigate the many thousands of acres now open to settlement.

The building up of the wonderfully rich Jerome and Cherry Creek mining districts should place upon the rich soil of the Verde Valley a much increased valuation during the next few years.

From the Santa Fe, Prescott & Phœnix Railway, Jerome is reached over the United Verde & Pacific Railway, a narrow-gauge line, which is said by engineers to be a most

UP IN THE MOUNTAINS — S. F. P. & P. R'Y.

wonderful feat of railroad construction. A trip to Jerome over this "corkscrew" road is one that should be taken by every tourist.

The Jerome section also has many cliff dwellings; through the Verde Valley there are thousands of these dwellings, many of which have never been visited by white men.

A twenty-mile drive through the Valley and you are at the Montezuma wells and the Cliff castles of a now extinct race; a few miles more over good roads and the Natural

Bridge is reached—one of the wonders of Arizona.

It is a fourteen-mile drive from Jerome to Beaver Creek, where the "Zulu" spring pours forth water that is claimed to be a sure cure for kidney trouble. Another short drive over first-class roads brings the tourist to the Grand Box Cañon which affords fine fishing and hunting.

INDEX.

	Page.
Announcement	Cover
Benefits of Climate Cure.	5
The Persia of America.	9
Arizona : a Winter Resort	11
A Perfect Winter Climate	22
Climate the Best Remedy	32
The Editor's Point of View	36
Sunnier Than the Riviera	37
Phœnix in a Nutshell	39
Hotel and Other Accommodations	42
As Seen by a Journalist	44
The Arizona Summer Climate	46
More About Summer Heat	49
Making a Living in Arizona	52
Prescott and Vicinity	57
Prescott Further Described	62
Castle Creek Hot Springs	64
The Mining Industry.	66
Jerome Mining Camp	68
Map of the Santa Fe Route	72
Outline Map of Salt River Valley	Cover

Be sure your tickets read over the

SANTA FE ROUTE (In connection with S. F. P. & P. R'y)

Via Albuquerque and Ash Fork.

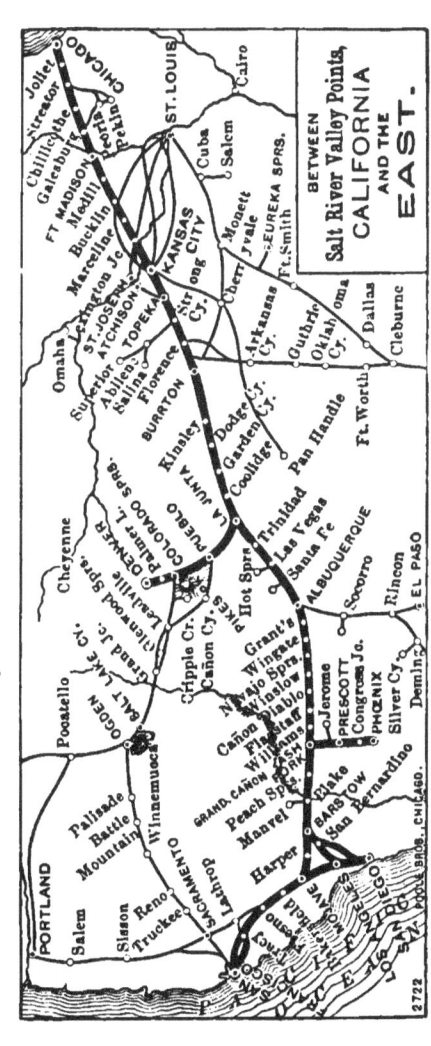

The Shortest Line to Southern Arizona.

The SALT RIVER VALLEY OF ARIZONA

Is 65 miles long and 15 to 25 miles wide.

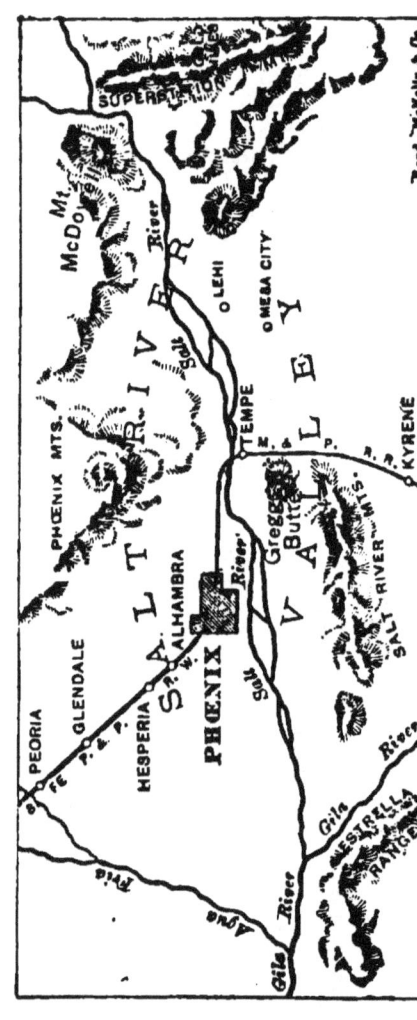

The irrigation system comprises over 400 miles of ditches, watering 350,000 acres.

www.ingramcontent.com/pod-product-compliance
Lightning Source LLC
Chambersburg PA
CBHW022150090426
42742CB00010B/1448